dog

SPEAK

Hubble & Hattie

recognising and understanding behaviour

www.hubbleandhattie.com

Hubble & Hattie

The Hubble & Hattie imprint was launched in 2009 and is named in memory of two very special Westies owned by Veloce's proprietors.
Since the first book, many more have been added to the list, all with the same underlying objective: to be of real benefit to the species they cover, at the same time promoting compassion, understanding and co-operation between all animals (including human ones!)
Hubble & Hattie is the home of a range of books that cover all-things animal, produced to the same high quality of content and presentation as our motoring books, and offering the same great value for money.

Photo credits
116 colour photos by Vivien Venzke/Kosmos.
Other colour photos by Melanie Grande/Kosmos (Pg 70, 71), Juniors Bildarchiv (Pg 12, 13, 57), Reinhard Tierfoto (Pg 46, 47); Christof Salata/Kosmos (Pg 69), Horst Streitferdt/Kosmos (Pg 59 top), Sabine Stuewer/Kosmos (Pg 21, 50, 58, 59 (bottom), 60); Sabine Stuewer (Pg 29, 32 (top and middle), 68), Jude Brooks (Pg 50, 56, 75, 76), Rod Grainger (Pg 59)

The publisher and author have designed this book to provide up-to-date information and advice regarding the subject matter covered, but cannot accept liability for any damages incurred to people, things, or assets, which may have occured as a result of following the advice in this book.

First published in English in March 2012 by Veloce Publishing Limited, Veloce House, Parkway Farm Business Park, Middle Farm Way, Poundbury, Dorchester, Dorset, DT1 3AR, England. Fax 01305 250479/e-mail info@hubbleandhattie.com www.hubbleandhattie.com/web
ISBN: 978-1-845843-84-7 UPC: 6-36847-04384-1. Original publication © 2009 Franckh-Kosmos Verlags-GmbH & Co KG, Stuttgart. © Christiane Blenski and Veloce Publishing 2012. All rights reserved. With the exception of quoting brief passages for the purpose of review, no part of this publication may be recorded, reproduced or transmitted by any means, including photocopying, without the written permission of Veloce Publishing Ltd. Throughout this book logos, model names and designations, etc, have been used for the purposes of identification, illustration and decoration. Such names are the property of the trademark holder as this is not an official publication.
Readers with ideas for books about animals, or animal-related topics, are invited to write to the editorial director of Veloce Publishing at the above address. British Library Cataloguing in Publication Data – A catalogue record for this book is available from the British Library. Typesetting, design and page make-up all by Veloce Publishing Ltd on Apple Mac.
Printed in India by Replika Press Pty

Acknowledgements

I would like to thank Renate Albrecht from the Dogs in Motion dog school, and Katy Schwania for the opportunity to photograph great teams in beautiful surroundings.

I would also like to thank photographer Vivien Venzke for such great team-work.

Special thanks go to my husband, Christian, my children, Bastian and Benita, and my wonderful Australian Sheepdog, Jaden, for everything!

Contents

Dog Speak: Quick translator!

I want to play!
This, the 'play bow,' is particularly seen in young dogs: the forelegs are placed flat on the floor, and the bottom is playfully stuck up in the air. The tail is held upright and wags enthusiastically, while the eyes dart cheekily back and forth. The dog is saying, 'I want to play! How about you?'

I was here!
Scent marking is not only common in male dogs: all dogs leave behind their individually-scented calling cards. Often the scent left by one dog will quickly be replaced by another. Dogs use this as a message, which says, 'I was here!' or 'I am the best!' or 'Suzie is on heat!' This scent-marking is like a noticeboard where dogs can leave announcements for each other.

I'm reading my p-mail!
For dogs, scents are like signposts everywhere that supply information about other dogs. A dog's nose is a highly specialised, high-performance tool. Using the Jacobson's organ (an auxiliary olfactory sense organ) dogs can effectively 'taste' a scent – especially if it contains pheromones.

Hello!
Whereas we use words to greet each other, dogs use body language and scent. Enthusiastic sniffing – mainly of the genitals and under the tail – is the equivalent of our handshake and a friendly, 'Hello, how are you?'

I'm happy!
Usually, a wagging tail means a happy dog. Or to be precise, a dog who is almost beside himself with excitement. Puppies also wag their tails to show adult dogs that they are ready for feeding. raising their tails to body height, and quickly wagging them back and forth.

Running with friends!
If a greeting goes well and both dogs are friendly, then the encounter can often develop into a game. Running together is a favourite sport with dogs: one runs away, and the other chases, then they swap over, and so the game goes on.

I'm completely harmless!
By lying on his back, this young dog is showing the Dogue de Bordeaux his throat and abdomen, in order to say, 'Dear stranger, I am tiny and completely harmless.' He then licks the corner of the dog's muzzle – another appeasement signal used between dogs.

I'm not sure about this ...
The black dog is clearly showing signs of uncertainty as he greets the bigger animal. He makes himself smaller, the tail is held low, the ears lie flat ,and the eyebrows are raised. His whole body language says he is very unsure about this situation.

We're ever so polite!
Standing sideways-on and circling around each other is all part of a polite ritual that dogs use to greet one another. They also signal peaceful intentions by using indirect eye contact. Dogs who are strangers do not usually look each other in the eye.

I'm hot!

This dog is hot and panting. His damp tongue cools the air he breathes in, and so helps reduce his body temperature. Panting doesn't necessarily mean he is thirsty, although dogs always need plenty of water in hot weather. During training, panting can mean, 'I am stressed and I feel out of my depth here!'

I don't like you!

It will be obvious when a dog is displaying very threatening behaviour. He will bare his teeth, draw himself up to make his body appear larger, shift his weight forward, bark and growl. He will also stare fixedly at the 'enemy.' This German Shepherd is clearly showing signs of aggression, and is ready to attack. Time to run for it!

How lovely to see you!

Wolf parents feed their puppies with pre-digested meat when they return from hunting, and their puppies greet them by jumping up and licking their muzzles to initiate this. Similarly, jumping up and nudging against your mouth is a sign that your dog is absolutely delighted to see you.

1

the
polite
dog

dog speak

Understanding body language:
Life as part of a pack

A pack of wolves is just like a family: each wolf has his own specific job to do, and most of the animals in the pack will get along well together when doing their duties.

Body language allows better understanding, and helps to avoid conflict

Dogs have descended from wolves, and wolves are pack animals. Although this is a well-known fact, have you ever considered how a pack actually functions?

Pack life

Living in a pack has many advantages. Wolves can defend themselves more easily in a pack than as individuals, and they are more likely to catch bigger prey, working together to hunt strategically. This means more food for each wolf, and healthier puppies who can go on to breed successfully themselves.

As in any family, each pack member has their own particular strengths and their own role, and it's a joint effort to ease the burden of daily life. To help group life run smoothly, there is a hierarchy within the pack. It was once thought that within each pack there was just one Alpha wolf in charge, and that all the other wolves were subordinate to him. But actually, this is not quite true.

Getting along

To ensure that life in a pack goes well, the wolves must learn to understand one another. If there is disagreement over something such as food, or a mate, the animals would achieve very little by spending all day fighting about it. They

Wolves fooling around – young wolves play together and learn how to refine their body language in the process.

would waste valuable energy, which could be used for hunting, raising their young, or defending their territory. They have, therefore, developed an effective form of communication amongst themselves.

The dominant animal

The head of the pack is in charge, but this does not necessarily mean he is the best fighter. On the contrary, he is extremely wary of conflict, and is a very good communicator. He may also have special hunting skills, being able to assess situations more effectively than the others. This makes the pack feel more secure. They recognise him as the chief, and happily follow his lead.

Pack members use displays of submissive behaviour to signal their compliance: making themselves look smaller, licking his muzzle, wagging tails, and lying on their backs. They do this not because the Alpha male has forced them to submit, but of their own free will to show that they recognise who is in charge.

An exchange

Wolves and dogs use body language to communicate with each other.

Specific signals are used to convey each animal's intentions and emotions; instantly understood by the rest of the pack. Of course, this use of body language must constantly be fine-tuned, so that the animals do not 'talk at cross purposes.'

You're the boss: the subordinate animal uses appeasement signals to show that he recognises the ranking order; he does this of his own free will.

Dominance and sub-dominance

The concepts of dominance and sub-dominance characterise the behaviour between two animals over the course of several meetings. Obviously, one animal on its own cannot be dominant! To say, 'my dog is dominant,' is nonsense.

The greeting ritual:

Dogs getting to know one another

The ritual of the first meeting begins from afar

After first eye contact, they sniff and snuffle around in the immediate area.

When dogs 'speak,' they move their whole body to do so. Of course, dogs can bark, growl and howl, too, but in fact, seldom do so when communicating with other dogs. They are more likely to use their eyes, muzzle, teeth, ears, fur and varying positions of the tail – a combination of these techniques helps them to communicate clearly to other dogs. The more effectively a dog can express himself using body language (as well as understand the body language of others), the more pleasant each meeting is likely to be with unfamiliar dogs.

He must learn the vital importance of good manners and etiquette from the start.

From a distance

Contact between dogs often begins long before we humans have even noticed it. Dogs can perceive movement from a very long distance: they might fail to notice the hare crouching in the grass just yards away, but they can clearly see the zigzagging rabbit right over at the far end of the field.

Dogs are expert hunters; they can spot the tiniest movement, and evaluate what type of animal it is as it gets nearer. If it's a dog, he checks, 'Do I know this dog or is he a stranger?' and so the encounter has already begun, soon to be followed by the greeting ritual. Setting his sights firmly on the other dog, he sprints off towards him.

'For my next move, I will walk around to the side of you.'

After the sprint, and discovering that 'Yes, this dog is a stranger,' the ritual then becomes an exchange of courtesies, for which there are various options. Some dogs keep a safe distance, sniffing the ground here and there, while gradually edging towards the other dog. Polite, well-socialised dogs, with commanding body language and strong characters, won't head directly toward each other, but make their approach via a slight detour.

Other dogs may use the 'stop-and-go' method: after they have spotted the other dog, they lie down, then get up and walk a little further before lying down again, and so on. This polite yet confident approach also makes them appear smaller, signalling that their intentions are friendly.

Playing it safe

When meeting for the first time, dogs need to ensure they clearly determine their ranking, and avoid any potential confusion in their communications.

Misunderstandings often arise between dogs due to uncertainty, and this can lead to aggression. In fact, a fight between two dogs is more likely to be caused by mixed signals, rather than a clash of dominant characters. If their body language is unclear, one dog could send out ambiguous messages, which the other may misinterpret as a negative signal. So if you own a puppy, let him socialise with other good-natured dogs right from the word go. This will help him learn important lessons about canine etiquette, which he will need to draw on throughout his life.

After the dogs have slowly approached each other, the ritual of getting to know each other begins: first sniffing the head, and then the hindquarters.

dog speak

Saying 'Hello' without words:

Scent contact

At their first meeting, the Australian Shepherd (left) holds his tail high, to give him an air of authority. The Icelandic Sheepdog firstly sniffs the Australian Shepherd's head, and then places his head across the back and shoulders, asserting his dominance.

Having noticed each other from afar, edged a bit closer, and made it clear that they come in peace, finally the moment of greeting arrives – a 'Hello' without words, or rather, without barking. Dogs use a very precise ritual here, in the same way that we would offer a handshake.

The home straight

As they finally reach each other, the dogs don't stand face to face, but instead approach each other diagonally, ending up standing side by side.

Now the greeting ritual can begin. Firstly, the dogs begin to sniff one another – their equivalent of a handshake. They start with the most important areas: the head, the tummy, between the hind legs, and under the tail. Very young dogs, or dogs who are

rather unsure of the situation, will lie on their backs and display their abdomens during the sniffing ritual, to show their subordinance to the other dog.

What happens next?

While the dogs are checking out each other's scent (or perhaps the scent left on a nearby tree), they can begin establishing the pecking order. Sometimes this can be immediately obvious – for example, when a confident dog meets a more timid character, he will adopt an upright body posture to show who's boss, while his new acquaintance falls into line, making himself look smaller by lowering his body, and behaving passively. However, if both dogs are equally dominant types, then each must try to assert his superiority; tails aloft, strutting around each other, each attempting to place their head on the other's back. Will one dog concede? Will both tails stay high and wagging, or will one finally stop? The suspense is almost unbearable!

No eye contact!
Watching this ritual take place, you'll notice that dogs do not use direct eye contact whilst meeting and greeting. This is a vital rule in the dog world, and one we'd do well to be aware of. No matter how lovely a dog's eyes may be, try not to look directly into them, as this could be interpreted as a threat.

The exception
Podencos are sighthounds (which means they use their excellent eyesight more than their sense of smell for hunting). When meeting a stranger, sighthounds will actively seek out eye contact with them. Of course, this often leads to aggressive reactions from other dogs. So, watch out!

Tip
Observing your dog

Keep an eye on your dog when he's with another:
What is your dog doing?
What is his body language telling you?
How does the other dog react?
This way you can begin to learn what the dogs are 'saying', and be ready to intervene if there are any problems. But remember – no direct eye contact!

Don't mind me!
Try not to stare too obviously at the dogs while they are greeting each other, if you are curious or anxious to check things are going well. By stopping and staring, you may create a tension which the dogs will pick up on. Don't call to your dog while he is busy with his 'meet and greet' – you won't succeed in attracting his attention. Once the meeting looks to be nearing an end, walk on a bit faster, and call your dog in a friendly but firm tone. He should then come to you.

No longer strangers; now the fun and games can begin!

dog speak

Understanding each other:

I like you!

By the end of the ritualised greeting, sniffing and sussing out who is the most dominant, it will be apparent whether the dogs can be friends. If both dogs have decided that they like each other, it will be easy to see: upright tails wagging, ears back, and eyes moving from side to side. You can sense it immediately – these dogs are on the same wavelength.

side by side, and share toys or sticks. Physical closeness is very important in packs of dogs: if two dogs live together, they'll often rest side by side, and groom, nibble and lick each other's coats, strengthening the bond between them.

Dogs who are good friends not only tolerate physical closeness, they enjoy it, and are even prepared to share their toys with each other.

Physical proximity

If a dog has decided that he trusts another dog, it will be obvious in his body language. Dogs who are strangers will only sniff one another in a perfunctory fashion, whereas the better they know each another, the more enthusiastically they will sniff, walk

Playing with friends

Canine friends will of course enjoy playing together. Puppies and young dogs play together all the time, and adult dogs are partial to the odd game here and there, too. These can involve chasing, running, jostling, biting – a whole range of hunting and play-fighting behaviour. When dogs are in 'play' mode, their behaviour, facial

expressions and body language are strongly exaggerated, while the roles of the chaser and the chased switch continuously. Body language can look confusing – bared teeth and loud growls, combined with playful wiggling of the rear end, before finally erupting into the chase.

Always keep an eye on these games just in case a conflict arises. The mood may change quite suddenly – if you notice that the game slows, and the dogs freeze in position, it could be time for you to intervene. Or, if one dog is being continuously chased or bullied, it's no longer a game, he's not enjoying it, and you should remove him from the situation.

Playing is practice for real life

In his book, *The A-Z of Dog Behaviour*, ethologist and behavioural biologist Roger Abrantes points out that "playing is a serious pursuit" – not just an amusement. Indeed, canine play has two main functions: firstly, it's an opportunity for puppies to improve their communication skills, body language and social behaviour. As early experiences will shape them for life, it is important for puppies to mix with other young dogs, and hone their social skills. The second function of play, especially for older dogs, is the chance to try out new behaviours without serious consequences; a dog can test to see if a particular behaviour leads to the desired reaction, and then either use it in everyday life, or not bother with it again. This is typical trial and error learning behaviour for dogs; if his behaviour meets with success, he will repeat it.

During play, almost anything goes: jumping, running, chasing, and swapping roles.

Tip

Lie down and connect

Lie down next to your dog. This will create trust between you and your four-legged friend, strengthening the bond between the two of you – and relax you both, too!

Clear for all to see:

Sorry, not interested

The black dog is rather anxious, and tries hard to make friends, but the Mastiff remains oblivious and indifferent.

As well as friendliness and hostility, there is one more possible reaction amongst dogs – disinterest. Sometimes, two dogs don't have anything in particular against one another, but they simply don't bond, which may be because they just don't know what to make of each other. It may be different if they were to meet again on another day, when they might feel more sure of each other.

The first sign of disinterest
Usually, it all happens quite quickly: the dogs sniff each other, and then pointedly turn their heads away, gazing in opposite directions (a sign of politeness). One dog may then attempt a second or third contact, or an invitation to play. But, if he pushes his luck, he might receive growling or snapping in response. The other dog may then turn his body away, wandering off a little, as if to say, 'There is no way I am playing with you!' When you continue on with your dog, you'll notice that he keeps looking back at the other dog, and, if you let him, he may return to where the other dog was to try and pick up on his scent, even though

there was very little contact between them.

Different sizes

It's common to see very small and very large dogs who get on perfectly well together. Size, therefore, is not a factor in determining dogs' responses to each other, whether that's 'like,' 'dislike' or plain indifference. (Interestingly, it is not known for sure whether dogs are aware of their size at all in comparison to other dogs.)

If dogs choose not to play together, it is simply down to differences in their natures or behaviour styles. If one dog shows disinterest, then the rejected friend (usually the less confident of the two) will lie down without attempting to initiate play. The first dog will give a quick, cursory sniff, and continue on his way. Within the few seconds of this encounter, the disinterested dog has made himself clear. If you replayed this in slow motion, you'd see how the dominant dog places his head over the other's shoulders, or how a subordinate dog makes himself appear smaller (by lowering his body), and quickly looks away.

Dominance

In general, I try to avoid using the word 'dominance,' since the idea of a 'dominant' dog can often be detrimental to training him: " ... he is just so dominant, nothing can be done," or even worse, it may lead to abusive behaviour towards the dog: " ... he's so dominant, I must do whatever it takes to get through to him."

However, as the Austrian vet Sabine Schroll explains in the film *The A-Z of Dogs*: "Dominance is a characteristic within a relationship, and not a personality trait. No one dog is dominant, per se. A dog may behave in a dominant way – but only during conflicts where he is more likely to win. His behaviour depends on the relationship."

If you observe a meeting between two dogs, you will soon see that they decide between themselves who's in charge – and it will vary with each encounter. This hierarchy needs to be established quickly, as ambiguity can lead to conflict.

When dogs meet whilst on the lead, conflicts may arise because their movements are restricted, making it difficult for them to clearly express themselves with body language.

dog speak

Making the most of each encounter:

A meeting of different generations

Most well-socialised adult dogs are tolerant of puppies, but there are limits ...

The photos here show that adult dogs and puppies differ distinctly in temperament. While most adult dogs are happy to lie quietly, puppies will wriggle about, lick each others' muzzles, sniff one another, bite fur, and try to initiate games. The older dog regards them patiently, willing to tolerate their playful antics for the most part.

The fountain of youth

The Mastiff's body language (below) shows he is feeling calm, with ears down, and his tail and facial expression relaxed. If the puppies' playfulness goes too far, there are several different ways he might react:

- Get up and walk away
- Give a short growl and a threatening look
- A quick, gentle bite on the muzzle
- Push the puppy off

Older dogs who already know a puppy and belong to the same pack won't have any desire to hurt him. Studies of wild dogs and wolves show that when puppies emerge from their den after three weeks, and begin to play outside, the whole pack approaches them with curiosity. Cynologist and author Gunther Bloch summed up this reaction after observing a pack of feral dogs in Tuscany: "The puppies are like a fountain of youth for the adult dogs."

Nudging the nose and licking the muzzle is a greeting ritual that puppies use with adult dogs.

Licking the muzzle

In the wild, licking the muzzle is a kind of survival behaviour used by puppies to stimulate feeding. To familiarise a wolf puppy with eating prey (instead of their mother's milk), adult wolves feed them pre-digested meat straight from their own mouths. Echoing this instinctive behaviour, domestic puppies may sometimes lick and nuzzle at an adult dog's mouth. Later, this innate behaviour serves as a communication technique amongst adult dogs, this time with a different meaning: nudging and licking the muzzle becomes a signal of submission, telling the other dog, 'I am placid and harmless.' Often, a dog will even lick his own muzzle in an effort to appease his opponent.

Licking the coat

The mother instinctively licks her newborn puppies to stimulate their circulation and to clean them. Similarly, when they lick her, it stimulates the flow of milk, and in turn, the mother licks her puppies' abdomens to aid their digestion, and stimulate defecation.

Hunting skills

Dogs shake their prey, but never their puppies. The mother will only pick her puppies up by the scruff of the neck if she wants to remove them from danger. Puppies, however, use the 'neck-shake' in their games as a form of training: this is no more than the puppies biting and gently shaking each other by the scruff of the neck. In doing this, they are practising their hunting skills.

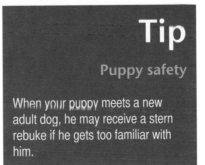

Tip
Puppy safety

When your puppy meets a new adult dog, he may receive a stern rebuke if he gets too familiar with him.

Neck biting and shaking is only used by puppies if they are role-playing a hunt. If the 'hunter' gets a little carried away, the 'prey' dog will fight back!

Paws for thought:

Do dogs learn from one another?

Dogs learn by trial and error, so if begging gets them what they want, they will do it again. Young dogs will happily copy older dogs, and unfortunately this includes the bad habits as well as the good!

A basic principle

Dogs learn from the reactions of their owners: a positive response to a certain behaviour will result in a repeat performance. If a dog pokes his head over the dining room table to steal a piece of cheese, and meets with success (ie eating the cheese and getting away with it), then he will do it again. However, if your dog jumps up onto the sofa for the first time and you

tell him off for it, your negative response will deter him from trying this again.

Is goofiness learnt by example?

No, dogs are not really observational learners. How much easier would training be if this were the case! You would only need to show your dog how perfectly another dog performs during an agility course, and your dog would copy him. However, young dogs do pick up the same bad habits as older dogs around them. Why? Because they copy the older dogs, following their example: 'Aha, now I have to go this way, and then slip between these two fence posts, and then I can spend an hour sniffing this field ...'

Dog dialect

At first, puppies grow up alongside their own kind. They share many similarities with their siblings: the shape of their ears, tails, facial expressions and fur. The mother, while bigger, also has the same characteristics. So the puppies only learn their own 'dialect' form of dog-speak.

Dog breeds vary widely in their physical characteristics. They may have:

- Big, floppy ears that just won't prick up, as with Beagles

- Permanently upright tails, like Icelandic Sheepdogs, for example
- Different-coloured coats, such as a Dalmatian's spots
- Fur that always stands on end along the back, for example, Rhodesian Ridgebacks
- Large, wide-open eyes, for example, Pugs
- A forward-leaning stance, as with sheep dogs

It's important to socialise your puppy with a variety of other breeds, so that he can learn to understand their different body language, rather than being able to read the signals from only one type of dog. Otherwise, he may remain ignorant of the significant characteristics of a particular breed of dog.

Curiosity is 'catching'

Dogs react strongly to the behaviour of other dogs, and this gives them opportunities to learn. If they notice that a dog has a particular interest in something, perhaps another dog or a person, this acts like a signal: they absolutely must go and see what all the fuss is about. Dog owners can easily test out this instinctive behaviour.

Squat down in the grass and stare fixedly at a particular place. As soon as your dog notices this, he will run over to see what exciting thing you have discovered (and you can then quickly deposit a little treat there for him!).

As a predator, he will also react to small movements and subtle noises, like the squeaking of a mouse. Loud noises won't be quite so fascinating to him.

Top: In mixed, supervised play groups, puppies can practice their social skills.

Above: 'What's that? Dogs are naturally curious, and love to discover exciting new things.

Tip

Body language

The better your dog is at running, balancing, fetching and swimming, the more expressive will be his body language, which means he will be able to communicate better.

Regular stroking and massage of your dog's entire body will help him become more at ease with expressing body language.

Full-on physical activity:

Puppies playing together

Yes, puppies do sleep a lot – but once awake, they're full of beans. Puppies can keep each other amused for hours, and always seem to have plenty to 'say' to one another. As they grow older, they learn to refine their communication skills, ready for adulthood.

Chase me!

A favourite game between puppies: one puppy charges off as if to say, 'Chase me!' and the second puppy runs after him. When he catches up, a playful tussle will ensue, and then a second bout of running, often with role-reversal. This game develops a puppy's co-ordination and fitness, preparing him for a real hunt.

One of the favourite games of a puppy is playing chase.

It's a good idea to take your puppy to a training class, so that he mixes with as many different dogs as possible, and learns the important skills he will need in adult life.

A little squeak

Two or more puppies together can get very rowdy, and puppies won't always tolerate being jostled about by their siblings. If they start getting a bit boisterous with each other, try making a small squeaking noise. This will often stop the noisy puppies in their tracks, because, to their ears, it's a clear signal, saying: 'That was too much!' They will soon learn to be more careful during the next game. In this way, we can sometimes actually 'speak' to our puppies. Similarly, if a puppy tries to bite your arm while you're holding him, emit a loud 'Arghhh!' sound and put him back down. This negative reaction will teach him to be more careful with his teeth, and will develop awareness about what he can and cannot bite.

Howling

In that first car ride to his new home, and during the first night with his new family pack, most puppies will howl. Finding themselves suddenly alone, they are trying to make contact with their mother and siblings. Howling goes way back to a puppy's wolf ancestry. In the wild, these calls can be heard

During play, the puppy is bitten too hard. He yowls ...

... and breaks off the game. The other dog learns from this reaction that his bite was too hard!

over a great distance, and they have several different functions, including guarding territory, and bringing the pack back together. For domestic dogs and puppies, howling is a clear signal of loneliness; they are missing the all-important togetherness of their familiar pack. So try not to leave your new puppy on his own, especially at night; put his basket by your bed to help reassure. And if your neighbour's dog joins in – which he may do as howling is rather contagious – this can also be quite reassuring for your puppy – though maybe not your neighbour ...

Tip

Milk teeth

Puppy teeth are sharp, and they grow very quickly, so give your puppy as many opportunities as possible to nibble and chew on safe objects, like a small beef bone or an old hand towel. The adult teeth begin to grow after about four months, and have fully developed by around seven months. At this stage, your home, your shoes and your hands should be safe from puppy chewing, and his awareness of what he can and cannot bite will have finally developed.

dog speak

dog speak

Familiarity: Good friends

Right: Friends who get on well may even share the same stick!

If two dogs have an ongoing friendship, then each time they see each other, their greeting won't be quite so drawn-out. Approaching each other no longer involves such a meticulous ritual, or carefully-defined boundaries. The dogs see each other, recognise who it is, and rush over to say 'Hi!'

However, if a third dog tries to join in, he may be given the cold shoulder.

An adventure for two

Dogs who know one another don't waste any time in beginning a game.

They run, frolic and tussle together and, of course, continuously switch roles to keep the game exciting.

Learning to share

Interestingly, some dogs can share their toys without any argument. Two dogs may run after the same toy which has been thrown for them, and will sometimes carry it back together, or one dog will happily let the other pick it up. However, occasionally, a toy thrown for both dogs can cause arguments. In case this happens, keep a supply of similar toys (balls, sticks or Kong® toys) at the ready to act as a distraction.

Two's company ...

When two dogs are playing happily together, a third may come along and interrupt the game, or try to join in; it can be interesting to observe this. The dogs will follow the usual greeting ritual, including sniffing all the significant body parts! The first two dogs may then react in different ways to this newcomer: one may pull back, while the other begins to either threaten, or excitedly greet, the newcomer. Sometimes, two dogs who have a close bond will not allow a third dog to join in, which can mean that the third dog will bark loudly in an attempt to get their attention.

Hormones

As puppyhood draws to an end, a change in behaviour can suddenly occur. Young dogs effectively become 'teenagers' when they reach puberty, and will begin to try and dominate bitches and other dogs. They no longer want to play the submissive role in games. Bitches suddenly begin to offer up their hindquarters, and are intensely sniffed. If they are on heat, they can send a dog wild with their seductive scent.

Mounting

When two or more dogs meet, you'll notice that one may try to mount the other, cling on with his forepaws, and begin to thrust with his hips. But is it sex? Not exactly. Bitches and castrated dogs also mount other dogs, no matter what sex they are. Mounting is a demonstration of power and dominance, used to emphasise a dog's status to the rest of the pack.

Mating

Of course, mounting is also a reproductive behaviour. If a bitch is on heat, she will give off a tantalising scent which a dog can detect from far away. Suddenly, no distance is too great, no fence too high, and the love-struck dog will go to any lengths to reach the female of his dreams.

The dog woos the female with a kind of canine foreplay: he frolics around her with a playful expression on his face, and licks her muzzle and ears. Once the female is standing still, the dog mounts her, and places his penis inside her. After a few forceful thrusts, he climbs off, but the pair remain stuck together because the swollen erectile tissue of the dog's penis prevents him from withdrawing. This is called 'the tie,' which can continue for up to 30 minutes.

Mounting has two main functions: to demonstrate dominance, and for mating, as shown here.

dog speak

Serious body language:

Appeasement signals

Appeasement signals

These are the essential building blocks of dog language: posture and small movements which placate other dogs, and signal friendly intentions.

The objective of appeasement gestures is to reduce tension

A variety of different signals

Appeasement signals may be grand gestures, or small, rapid movements. A dog will use them very early on in a new encounter, in order to fend off any possibility of a threat from another dog or person. These behaviours will also help him feel reassured if he is nervous in this meeting. The main goal is to show the new dog or person that he is friendly and means no harm. The signals include elements of the humble, puppy-like behaviour of an anxious dog.

No aggression

Pawing (nudging/touching with the paws), invitations to play (placing the forepaws flat on the ground with the

The little Weimaraner is quite clearly saying, 'Please don't hurt me!' by placing his ears flat against his head, making himself look smaller, and pressing his tail between his legs.

hindquarters in the air – the 'play bow'), and nuzzling around the mouth are all behaviours that go back to puppyhood, as if the dog is appealing to the paternal instincts of his opponent.

Other appeasement signals are the opposite of aggressive body language: keeping the mouth closed (instead of teeth baring), averting the gaze (instead of staring fixedly), and turning around and showing the hindquarters. A sure sign of appeasement is when a dog turns his bottom towards other dogs – or to you!

Displacement activity

A dog might also avoid conflict by using displacement activities, such as wandering off and digging, sniffing the ground, or turning away and scratching himself (see panel, right). By doing this, the dog breaks off the conversation, and shows his opponent that he is not interested in having a fight.

Careful observation

The list above right demonstrates the importance of understanding your dog's body language when he interacts with other dogs (and you).

Of course, a dog might yawn because he's tired, but it's worth bearing in mind that if he yawns during a training class, for example, this might actually be because he's feeling uncomfortable. Similarly, dogs often scratch their throats in unfamiliar situations – not necessarily because it's itchy. A dog may use this signal to ease any tension, especially if lots of people are standing around him and looking at him, which he will interpret as being stared at threateningly (see photo, right).

Displacement activities

- Turning the head (or the entire body) away
- Standing to one side and displaying the hindquarters
- Squinting or narrowing the eyes
- Quickly licking the nose
- Making himself appear physically smaller, eg by lying on his back
- Raising the paws, or scratching
- Yawning or panting
- Slow motion movements to counteract a flurry of activity
- Sniffing, in tense situations
- Other movements such as sitting down, lying down or wagging the tail

For the kids:

How to make your dog happy!

Lie down next to me!

You may want to show your dog how much you love him with a hug. But your dog is not a human being – it can actually be quite threatening for him when you bend over him, and throw your arms around him. He may react to the hug in a rather scared way. In any case, he will probably feel uncomfortable. Try this instead: lie down on the floor, and your dog will come to you and lie close next to you. This type of contact will help you form an even closer bond with your dog!

Look – no hands!

When you are playing with your friends, your dog may run behind you, and try to catch you by biting at your clothes. When dogs play chase with you (and most have a hunting instinct which they simply must follow!) they can only use their teeth to grab you with. So they may bite into your clothes, and possibly even hurt you by mistake. Try throwing your dog a ball when you want to play with him, so he will chase that instead of chasing you!

I need rules!

Playing with your dog can be great fun. But if it is always a matter of doing what the dog wants to do, there can be confusion over who is in charge. Remember that you are in charge! Try this instead: throw your dog a toy and then look the other way. See how he reacts. And make sure it's you who begins and ends every game. Tidy all the toys away at the end of each game as this shows your dog that all the toys belong to you.

Take it easy!

When you are out walking your dog, he will notice all the other dogs, people and traffic around you, or even a cat scurrying away in the distance. When you are absorbed in your own thoughts, it's a shock when your dog suddenly pulls hard on the lead, and you nearly fall over! Instead, try to keep an eye on what's going on around you, and stay alert to any possible distractions ahead. Hold the lead firmly in your hand, but allow a bit of slack in it, so you can let it out when necessary, rather than stumble and fall if your dog suddenly pulls forward!

Try these tips for a happy dog:

- Fixed meal times, and time to play everyday
- Practising tricks and training, making sure you reward your dog with treats
- Not pulling on his lead when he is walking alongside you, but using a nice loose lead
- Allow him his quiet time – dogs need much more sleep than people do
- No screaming or loud music – dogs' hearing is far more sensitive than ours
- Time together – being with you is your dog's favourite hobby

At a glance:

Keeping the peace

Playing

Almost anything goes in play, and dogs can use it to check out each other. Through playing, puppies learn about canine communication.

For example, an invitation to play can diffuse a tense situation, and a friendly game is clearly signalled by non-aggressive body language. However, if one dog tries to playfully bite the other, but ends up biting too hard, the other dog may howl and break off the game. If the game becomes too wild, or one of the dogs is being bullied, they may need your help to stop the game.

Sniffing around

Dogs tend not to run straight towards each other, but follow the polite etiquette of a new meeting, for example casually sniffing at the ground near each other, or lying down, getting up, walking closer, and then lying down again. When they do run towards one another, it's in a rather zigzagging manner.

Allow your dog to meet other dogs off the lead so he has a chance to practice his dog-speak.

Tips for meeting another dog

- *If possible (as agreed by the dog owners and in a safe environment), take the leads off*
- *Always keep the dogs within view*
- *Stay alert: any sudden excitement could distract your dog*
- *Keep going: walk on rather than just staring at the dogs*
- *Protect your puppy: when he meets another dog, crouch down next to him and reassure him*
- *Never assume that other dogs will be friendly towards your puppy. Only socialise him with good-natured dogs*
- *Hide your dog's toy in your pocket to avoid arguments over it with the other dog*

Sniffing each other

Sniffing is a greeting ritual that's also used by dogs who are already good friends. It allows them to collect information about the other dog's age, sex and social status, and they can decide who is in charge during this process.

Turning the head away

Turning the head away is one of many appeasement signals used to pacify a threatening situation. The dog uses head-turning and glancing away to show that he comes in peace. It can also be a sign that the two dogs are not particularly interested in one another. Firstly, they look away, then they walk towards each other, but keeping their distance.

Lying down

If a dog rolls onto his back and allows himself to be inspected without resisting or moving at all, then he is expressing peaceful intentions.

But, if he struggles, fidgets, kicks or snaps in this position, then he is not submitting, and any dog standing over him will probably growl and snap in response.

Wagging the tail

Everyone knows that if a dog is wagging his tail, he is being friendly. But take a closer look at the body language as a whole. A wagging tail is only a display of friendliness when it is held at around body height, and is wagging rapidly. If it's held high, and wagging slowly, keep your distance ... Do explain to children that a wagging tail can mean different things – misunderstanding this behaviour can lead to disastrous consequences!

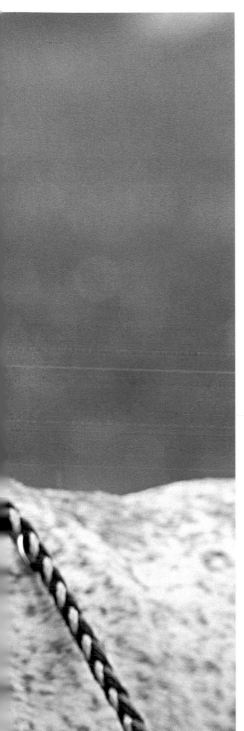

2

the aggressive dog

dog speak

Misunderstandings: The tail says it all

'Here I come!' This dog looks like he's got it all sorted. He stands upright to emphasise his large size, tail pointed upward, and, with an imposing air, confidently approaches the other dogs.

Nature has blessed the dog with a dazzling array of ways to communicate – especially impressive with some breeds such as the Collie, Beagle, and Australian Sheepdog. The tail – a versatile and expressive tool – is vitally important for the dog to clearly speak his language. Unfortunately, for some, selective breeding has rendered their tail almost useless for communication, permanently pointing upwards or perhaps even docked. Dogs still manage to speak their language, in spite of this impairment, but it is the tail that emphasises any point they are trying to make with the rest of their body.

Hostile or happy?

To show he is happy, a dog will wag his tail at body height, but an agitated or threatening dog will carry his tail somewhat differently. Roger Abrantes, author of US bestsellers, *Dog Language – An Encyclopedia of Canine Behaviour* and *The Evolution Of Canine Social Behaviour*, notes: "A tail that wags slowly and at a very low height can be a warning of attack." So the position of the tail, in addition to all the other body language signals, is crucial for understanding a dog's message. Threatening or aggressive behaviour is signalled by the slow, deliberately wagging tail, combined with the following gestures and behaviours:

- Staring
- A provocative look
- Wrinkling the nose, or baring the teeth
- A deep, throaty growl
- Clearly audible snapping of the jaw
- Stiff or tense posture
- Fur standing on end around the neck area

Meeting a threatening dog

If you see one or more of these signs from an approaching dog, do not let your dog off the lead under any

circumstances. Nor should you let the dogs 'sort it out between themselves.' You do not know whether the dog simply wants to express his aggression with no intention to attack, or is actually out of control and liable to cause physical harm. Keep up a brisk pace, but don't rush; try to act casually to help keep the atmosphere relaxed; head off in a different direction, or turn into a nearby driveway if possible.

This may seem like running away, but in fact you are simply protecting your dog from a potentially dangerous situation. It will also show your dog that the next time a hostile hound crosses his path, he can rest assured you will keep him safe from harm. In addition, avoiding any nasty encounters will reduce the chances of your own dog threatening other, friendly dogs. Otherwise, unprovoked aggression (caused by anxiety) can become a vicious circle, and one that is difficult to break.

Another message from the tail

To assert their dominance, very confident dogs carry their tail high in the air, and hold it still. You may notice that when a dog like this meets

a new canine acquaintance, he receives a rather wary, unfriendly or tense response; such self-assured body language often irritates other dogs, and can lead to confrontation.

Showing fear

The tail may also be used to express anxiety or fear. When a dog feels afraid, his tail will drop down between his legs, and curve under the abdomen. He will also try to make himself appear smaller, press his ears flat against his head, widen his eyes, and glance about nervously.

Tail under the tummy

If a dog feels particularly unsure when a strange dog approaches him, not only will he try to appear smaller, he may also lie on his back with his tail between his legs. Ethologist Roger Abrantes explains: "Puppies and young dogs wag their tails, even when the tail is clamped between their legs or they are lying on their back. By doing this, they are showing complete submission. By wagging their tail, they are also spreading their scent in an attempt to appeal to the maternal feelings of the adult dogs." So this behaviour is used as an appeasement signal. Young dogs may also urinate if they are unsure or anxious.

The sensible thing to do if you meet another dog who seems unpleasant or threatening is to quickly retreat.

The tail pulled in under the tummy, and his general body language show how uneasy this young dog is feeling.

Bad behaviour:

Look me in the eye, little one!

Long distance meeting: the black dog has already fixed his gaze intently on the dog in the far distance.

If you stare intently at a dog, he is likely to consider this threatening

When two people are in love, they spend hours looking into each other's eyes, showing their mutual appreciation and intimacy. Not looking someone in the eye is regarded wth suspicion, suggesting that they are hiding something.

Of course, with dogs it's a completely different story, just one example of the many differences between dogs and people. There are a few exceptions to this particular rule – Podencos, for example – but generally, direct eye contact can quickly lead to canine conflict.

Aggression at a glance

A fixed stare is often an early warning sign of attack, and can be made from a long distance. Dogs have excellent long-range vision – a far-off moving object will cause them to stop and stare. If the movement turns out to be another dog, then the usual greeting ritual will be put on hold. When this stranger approaches the dog who spotted him from afar, his body language will be noted and assessed: is he staring back, or just glancing around, turning away, or giving off other appeasement signals?

Additional factors

Whether or not a fight could break out depends on two things. Firstly, how aggressive is the threatening-looking dog? And secondly, how well does he understand the other dog's body language? There are also other factors at play: are the dogs meeting off the lead, so can roam freely out of one another's way? Or are they meeting in a narrow path on a short lead? If so, the dogs won't be able to fully express their canine body language, or maintain that all-important personal space.

The purpose of the eyebrows

For most dogs, the eyebrows are a tell-tale sign of their feelings. They are often raised, lending the dog a rather regal facial expression, and reinforcing

Dogs are more likely to resort to aggression if they are on a lead rather than able to roam freely, since they can neither fully express themselves using their body language, nor can they maintain their own personal space.

whatever the eyes are saying. If the dog's eyes widen, the eyebrows will be raised; if they narrow the eyebrows seem to disappear.

Read my lips

A dog can pull his lips back or push them forward so that his teeth are either completely exposed or covered.

The self assured dog will bare only his front teeth when he growls, pushing his muzzle forward, whereas a dog who feels anxious pulls his entire muzzle back when he growls, revealing his full set of teeth. Be careful, though: nervous dogs are just as likely to bite as dogs who are more assertive.

Dogs laugh, too

The human laugh might actually sound quite threatening to a dog's ears. However, some studies claim that dogs who are very in tune with their owners may actually mimic human laughter, and adopt a 'laughing' facial expression – though this is reserved for people only, never another dog.

Left: Eyebrows enhance the facial expressions.
Right: Even dogs can 'laugh'!

dog speak

A clear advantage:

Speaking with the ears

A dog's prominent ears are a great asset. Rotating in all directions, with no fewer than 17 muscles, they allow a dog to hear exactly where a particular sound is coming from. The canine ear can detect very high frequency noises which we cannot hear at all. Humans can hear soundwaves of up to 20,000 oscillations per second, whereas dogs can pick up at least 40,000 (sometimes as much as 100,000). Interestingly, however, dogs barely utter any sound at all when they greet one another, since so much information is transmitted through scent instead.

Dogs' ears are very flexible, highly specialised tools essential for canine communication

The position of the ears

Besides their auditory function, ears are vital for canine communication. A dog's ears frame the full picture of whatever he's saying: if his tail is pulled under, head low and eyes averted, then his ears will lie flat on his head. Or to emphasise a more threatening look, his ears will stand upright and point forward – he may also do this when merely curious about something, perhaps a mouse running across the lawn, or the sound of dinner being served!

When a dog's aggressive behaviour is prompted by uncertainty and anxiety, you can see that the muzzle is pulled back tightly as he growls, and the ears lie flat on the head. Always take into account all of the body language shown, therefore, as well as what is going on

Pricked ears are a more expressive means of communication than, for example, the ears of a Cocker Spaniel, which lie flat against the head.

around the dog, in order to assess the situation properly.

On the move

When dogs greet one another, you'll often see their ears moving forward, backward and from side to side. Out on a walk, at home, or in the garden, your dog's ears won't stay still for long – soon enough he'll be looking around, suddenly alerted to a new sound or another dog to greet. Then he will stand tall, proudly drawing himself up and make himself look bigger, perhaps by shifting his weight onto his front paws, and pushing his chest forward. If you notice him taking this stance during a walk when a new dog is approaching, it's time to move on! Just keep walking – this will clear any tension, and once the dogs can no longer see each other, he will stop trying to look imposing.

Between the ears

A dog's brain differs distinctly from the human brain as it does not have the 'Broca's area' which, in a human brain, is the centre for understanding and speaking a language. However, dogs learn fast, and are able to quickly make a connection between a certain behaviour and the consequence which follows.

Dogs do have their own language – body language – spoken from the end of the tail to the tips of the ears, which develops further with each new experience they encounter.

These two feel particularly uncomfortable because they are closer to one another than they would like to be. This is why the ears are down and the whole body language says, 'I don't want any trouble, buddy!'

Raised hackles:
The neck and back

Are you familiar with the 'brush'? No, not the wiry grooming brush you use on your dog, or even the brush you might use on your hair. This particular 'brush' is created by all the hairs on a dog's neck standing on end. This action also belongs to the repertoire of dog-speak ... but what is it saying?

🐕 **The animal with the highest status is entitled to more resources such as food and territory, and also allegiance from the rest of the pack**

The brush

As its name suggests, the Rhodesian Ridgeback has a raised ridge of fur along the back, running in the opposite direction to the rest of the coat. But, for most dog breeds, this fur will only stand on end if they are upset or disturbed in some way. This is the body language of a dog trying to make himself appear larger and therefore more threatening. The effect is emphasised by the coat being much darker around the neck area. The hair returns to its normal position once any conflict has passed. It may also rise as a sign of uncertainty and stress; so again, it's wise to take the whole situation into account.

This Jack Russell's fur is standing up on the back of his neck and he is scent-marking. He refuses to acknowledge the ranking of the Icelandic Sheepdog. But the latter dog asserts his status by standing over the Jack Russell, with his fur up on end, and placing one paw on him.

Making an impression

'Hey, I'm the strongest!' Dogs try to assert their dominance just as humans do. Normally, this dialogue happens between dogs who are of a similar social ranking, since a high-ranking dog has no need to explain his obviously

Laying down the paw

If his assertion of dominance hasn't worked so far, there are a few more signals a dog can try. Can he get away with placing a paw on the other dog's body, or resting his head along the back? He can also try the so-called 'T-position' – a ranking behaviour, whereby a domineering dog will stand sideways in front of the other dog, thereby reducing his opponent's room for manoeuvre.

The question of who is dominant will eventually be resolved by the other dog's reaction, when and if he finally backs down and concedes. If not, then further 'discussions' between the pair will ensue ...

Mounting

As previously mentioned (see page 29), mounting is another demonstration of dominance. A high-ranking animal will mount the other dog and hang on with his forepaws, quite clearly invading his personal space. This behaviour also sends out a clear message of superiority to any other dogs who may happen to be watching from a distance.

Why is this necessary?

In a pack, the animal with the highest status is entitled to more of everything: food, territory and even toys (in a domestic setting), as well as a greater chance of reproducing, and other social advantages such as being groomed and shown allegiance by other members of the pack.

The Icelandic Sheepdog stands sideways, in the way of the other dog. With his tail raised high in the air, he then places his head over the other dog's shoulders.

Lying down: what it means

A dog who lies on his back, displaying the abdomen, is usually appeasing and submitting to the other dog, but not always. If he also fidgets and snaps at the dog standing over him, this is anything but submissive behaviour, and the other dog may growl and bite back in response.

superior status to anyone. If, however, the pecking order is unclear, a dog may draw himself up to appear bigger, and stalk toward the new dog in his path with a fixed stare and tail raised high; generally trying to make it quite plain who is in charge. Scent-marking is also part of this dominant behaviour.

And yet in spite of all these attempts at dominance, the other dog may refuse to accept his authority, leaving the question: who is the higher ranking of the two?

In the midst of conflict:

Tense moments

The Husky below may appear threatening, but the second image shows he is actually slightly unsure of himself. The tongue and the way the muzzle is pulled back give him away, while the dog on the right reveals fewer teeth, and fixes his gaze.

Have a good look at your dog's teeth – you will see how sharp they are. Your dog has an unbelievably strong jaw, but how often do dogs actually get involved in fights to the point where their bite will draw blood? The truth is, even if you walk in areas where dogs socialise a lot, you're unlikely to ever witness more than a couple of serious fights.

Avoiding a fight

Thankfully, it's rare for a fight to result in a visit to the vet; the sophisticated language of dog-speak helps to resolve

upsets without the need for a nasty bite. Appeasement signals, like a white flag of surrender, will usually prevent things from getting too ugly – but what if a tense situation does escalate, and two dogs start to threaten each other, hackles raised, ears and tail upright, teeth bared, and eyes staring directly at the enemy?

In the thick of it!

The first sign of a threat can very quickly develop into an actual attack. Even a well-socialised dog will quickly

run through his repetoire of threatening behaviour until the point of no return. Then the bite will come with almost no warning. If you could watch the attack played back in slow-motion, you'd be able to see the exact moment where one dog raises his head higher, and the other pulls his ears back to avoid injury, just before the fight breaks out. When witnessing a fight like this at normal speed, it is almost impossible to work out exactly what happened: it's lightening-quick. The dogs become entangled in an almost inseparable bundle, trying to bite each other's throats. This is accompanied by a mixture of shrieking and barking: according to animal behaviourist Anders Hallgren, the dogs attempt to frighten one another with this bellowing so that they " ... avoid sinking their teeth in, or if they do bite, it is usually not too forcibly. This behaviour functions more as a theatrical display, instead of using real full-blown aggression." For this reason, there are often no injuries when the two are broken apart. This type of fighting is strongly ritualised, just like canine greetings, and is a sort of show-fighting.

Drawing blood

In contrast to this show-fight, a real dog fight involves attempts to seriously injure or kill one another. This battle is less noisy and more purposeful; it will be fought to the very end, until there is a clear winner and a loser.

The stakes are high. Amongst wolves, the defeated fighter must leave the pack for good. Life as a lone wolf is not easy: he must look out for danger alone, hunt alone, and will have no reproductive partners. For domestic dogs, a serious fight would also mean they could no longer live together,

since the victor would not tolerate his weaker opponent. Luckily, this is a rare occurrence.

The best way to intervene

Have an umbrella to hand: push the closed umbrella firmly between the dogs, or open it up suddenly right next to them. This may momentarily shock the dogs and distract them from their fight.
Keep a bottle of water handy, as a sudden unexpected shower may cool any hot tempers.
Some owners grab hold of their dogs by the hind legs, lift the dog in the air and pull him away from his opponent. Practice this in advance, though, as it should be done in one smooth movement during the fight.

A ritualised fight between two bitches.

Paws for thought:

On the lead and behind the fence

Aggression

Dogs can be noticeably more aggressive when they are on a lead. Canine body language becomes difficult to express, as movement is too confined to make it clear to another dog that they mean no harm. Sometimes, dogs may be forced to invade each other's personal space (in an alleyway, for example), and be unable to avoid each other. This forced proximity can lead to conflict.

So what can you do about this? Either let the lead right out as far as you can, or walk around the other dog, giving it as much space as possible.

Protect your dog

Many dogs seem to feel more secure when they are on a lead and with their owner, but if you pull on the lead if you think you've spotted a potential canine threat ahead, you may be making any tension worse, before your dog has decided for himself whether the approaching dog is friend or foe.

Stay relaxed, breathe slowly and, if in doubt, turn around or otherwise avoid the oncoming dog. Your dog does not need to greet every single fellow canine he sees!

Behind the fence

If a dog is barking behind a fence or gate as you pass by, do not let your own dog near it. In this excitable situation, and with a barrier between them, it won't be possible for the two dogs to communicate properly using their dog-speak. It's interesting to note that when the same two dogs meet off the lead, and with plenty of space to move around, they will probably get on very well indeed.

Get in between

Here's some training advice for dealing with dogs who react to the boisterous 'fence-barker.' Firstly, position yourself between the fence and the dog. Then practice walking past a 'quiet' fence which is not guarded by a dog. Later, when you come to a fence with a barking dog behind it, quickly whip out some tasty dog treats. This will get your dog's attention and encourage him to stay quiet. At first, increase the distance between your dog and the fence by leading him over to the other side of the road. If your dog gets past the fence and ignores the other dog, then next time you pass it you can gradually walk closer to the fence.

Watchdog

It's entirely unnecessary for a dog to defend 'his' garden by barking at the fence or gate. If you don't want your dog to be on constant alert, get him into the habit of keeping quiet at the fence. You can work toward this by not letting your dog in the garden by himself. If you are on the patio, he should be near you, and on the lead (in a cool, shady area). If you go inside, your dog should go, too. If he barks at the fence, go over to him without saying a word, and lead

him away from the fence. If he walks up to the fence now and again to have a good bark, put him back on his lead so that you can lead him away each time before he begins to bark.

Also, be sure to give your dog lots of exercise to tire him physically and mentally, and keep him happy so that he's less likely to become agitated, and is too tired to bother getting up to bark through the fence anyway!

At a glance:

Problem solving

a) Staring

If your dog has fixed his gaze on another dog, this means he will hold the stare and not let the other dog out of his sight. You can sense immediately the tension that is building.

a) Solution

Stand in the way of your dog's fixed stare. This will diffuse the tension, and also help to relax the approaching dog. The more relaxed both dogs are, the better they will communicate and avoid conflict by using appeasement signals, and the right body language.

Extra tip

Instruct your dog to 'sit!' whilst you are standing in front of him. It will break the tension, and, with his tail flat on the ground, your dog will appear more friendly to the other dog. This command will also draw your dog's attention to you, and away from the other dog.

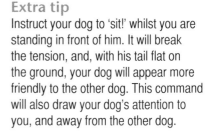

b) Barking

Barking can carry different meanings: a warning, a threat, uncertainty, or perhaps a cry for attention. What should you do when your dog begins to bark?

b) Solution

Keep as calm as possible, both inwardly and outwardly, remaining quiet and still. A dog does not understand your words, but he can respond very well to the sound of your voice. If you try to calm the barking dog with gentle words, to him this could sound like praise, so he will keep on barking.

If you scold him, he will think you are joining in, and will just get louder. The best way to stop him barking is to distract him by making a sudden, loud, unexpected noise, such as 'Psssst!'

You should remain quiet when your dog growls. If you forbid him from growling, he might not be able to warn you about something important.

c) Pulling on the lead

An irritated, aggressive dog does not tug gently at the lead. He pulls and yanks and drags you in the direction of another dog with all of his strength and determination.

c) Solution

Hold the lead very firmly, and get ready for the onslaught: move closer to your dog and hold him by the collar, or the collar-end of the lead. For dogs who

Tension-busters

- *Always: stay quiet and calm, assertively removing your dog from any situation that is uncomfortable for him*
- *On the lead: turn and walk in the opposite direction, or walk around the other dog*
- *Beforehand: practice getting your dog's attention, perhaps by using a special 'signal' word, or a game*
- *Try an ice-breaker: a non-threatening distraction like clapping your hands, or casually whistling, may help to dispel any tension during a lead-free greeting between two dogs.*

often tug at the lead, a harness can be a better alternative to a collar. This is a much kinder way of controlling your dog, in any case, and will protect his neck and make it easier for you to hold on to him. Lead your dog in the other direction with purposeful strides. If that's not possible, have him sit until the other dog has gone by. Don't pay attention to the other dog but try and come up with a tasty treat or a toy with which to hold your dog's attention until the other dog has passed by.

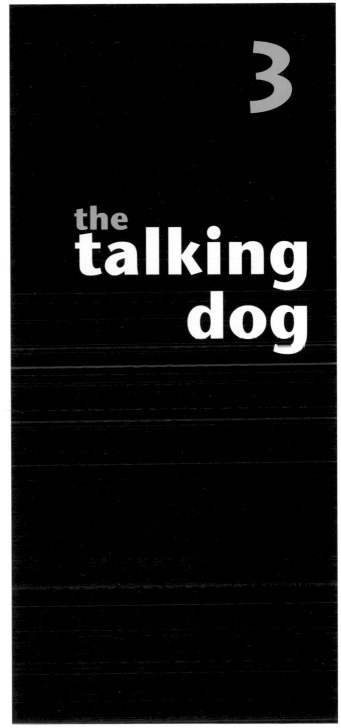

3

the talking dog

His whole body says: "Hello, human!"

Jumping up and licking the muzzle is a very friendly gesture between dogs, but not every person appreciates this!

Although a greeting between dogs is usually polite, when a dog greets his human family, he is more direct and boisterous. He will come running, eagerly wagging his tail, sometimes so energetically that his whole body wags, too. Depending on age and training, he will jump as high as he can and begin the sniff check to see what interesting smells his human has brought in with him. His ears lie flat and the head is held low in what almost seems like appeasement behaviour, even though he is evidently feeling confident.

In a nutshell, your dog is using his whole body to show that he is absolutely delighted to see you. He is saying, "Hello, human!"

Getting to know you

When a dog meets a stranger, he will revert to his normal greeting rituals. If he does not feel threatened by the stranger, then he will greet them quite warmly. To help ensure this happens when you meet a dog for the first time, try these tips:

Cuddling or hugging is a friendly greeting between people. Not every dog will tolerate being hugged, although, for some, being held firmly is reassuring.

- Don't look him directly in the eye. Look away, turning your head to one side
- Don't bend over him. Instead, squat down and wait for him to come to you
- Don't stroke his head: stroke his side or back
- Don't say anything at first. Wait until he is standing next to you, and then speak to him in a friendly tone

- Don't laugh. Try not to show your teeth, and keep your facial expression relaxed

By following these guidelines, a dog will feel safer when he approaches you, and will probably be excited to see you. A dog's memory is as good as an elephant's when it comes to positive and enjoyable experiences, so show him your best side!

Turn away

If you don't want to make contact with a particular dog, turn away as he runs toward you, and don't say anything. This body language will signal your disinterest. Try to keep a relaxed posture as you stand, so that you don't look tense. The dog should then lose interest in you and continue on his way.

Turning away like this is a very clear statement in body language terms, understood by both human and dog. It can be used if a dog rushes toward you looking rather too energetic and boisterous. If he tries to jump up at you, turn around and remain standing straight. Jumping up and licking the face is friendly behaviour, with its basis in evolution (wolf puppies are fed meat directly from their parent's mouth after hunting).

This person's body language says, 'Sorry: not interested!', which a dog will immediately understand.

Even so, ensure you train your dog not to jump up at visitors, since not everyone appreciates this.

Try this: Crouch down and wait until the dog comes to you. Then you can tickle his back. Both dogs and people love this!

Holding out your hand

The old advice: Hold out your hand for a dog to sniff
The new advice: Be careful! Your hand may be mistaken for a toy or tasty treat if the dog is in high spirits

dog speak

Go, go, go ...
Play a game with me, please!

Dogs can sleep for up to around 16 hours a day, so are generally awake for around eight hours. For how many hours of these eight is your dog active? Two or three hours of walking? Great. That leaves five hours. A little food, a little sniffing around in the garden, a journey or two in the car, gnawing at a bone, trotting around the dinner table and hanging out in the kitchen in the hope of a titbit ... that still leaves a couple of spare hours when he could get bored ... which is when he might just say to you, 'Come and play a game with me!'

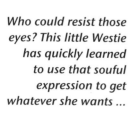

Dogs sleep for a lot of the day, but when they are awake, they're usually ready for action!

Who could resist those eyes? This little Westie has quickly learned to use that soulful expression to get whatever she wants ...

The paw on the knee

This is an unmistakable signal from a dog: 'I want something!' With one paw placed on your knee, and his gaze directed straight at you, his whole body is saying it loud and clear. There are several things he may be after: food, a walk, or maybe to play a game with you. So what should you do for him? Nothing, if you don't want to play then.

But if you are happy to do as he asks, fetch a little treat and have him sit first before giving it to him, reminding him that you are in charge.

Laying his head on your knee

If a dog (of any breed) lays his head on your knee, and fixes you with a melting gaze, it simply means he wants a nice scratch behind the ears. Should you oblige? Sure! However, don't say yes every time he asks you, and remember that if you do stroke him, it's your decision when to stop. When you say 'enough,' that's it, even if your dog wants more.

Growling

If a dog begins to growl after you've been stroking him for a while, then stop and leave him in peace. The growling means, 'I'm fed up, now.' If you don't take heed of the growling, he might end up snapping at you.

Bringing you a toy

Your dog strolls into the house carrying a ball in his mouth. What should you do? Either ignore him completely, or encourage him to drop it. Then, without a word, put it in your pocket. Some dogs will want to play endlessly, and annoy everyone with their constant requests for the ball to be thrown for them. Follow

Just like Lassie!

Dogs are the only type of companion animal who will ask a person to do something for them. If his favourite ball has rolled under the sideboard, your dog will rush over to you and, by barking and running back and forth, ask you to retrieve the ball for him. Dogs are unique in this respect; making communication between human and dog extra special.

this rule: the toy belongs to you, and it's you who starts and ends every game!

Nudging

Nudging is often used if a dog wants a treat. This behaviour goes back to puppyhood, when seeking milk from the mother. Between adult dogs, it is an appeasement signal. And when a dog nudges a person, it's either a prompt for something like a treat or his dinner, or he's simply looking for your attention. You might not mind being nudged sometimes, but don't encourage it by then offering him a reward such as a treat.

'Can you throw the ball for me, please?'

dog speak

Beware: Bad mood!

All dogs have their own individual temperament, partly influenced of course by their breed, personality and age. It also depends on whatever mood they happen to be in. Your dog may simply be feeling grumpy one day, making him oversensitive and short-tempered; and the next day he'll be cheerful again. So keep in mind that it can vary from day to day, and check his body language for telling signs.

Clipping claws etc

Unless you are proficient at this, it's best to get your dog's coat clipped by a professional. Doing it yourself may not be appreciated by your dog. He may well object, especially if you try to clip his claws, and is likely to become fearful. His first instinct will be to struggle and escape. If he can't get away because you are holding him, he will show his discomfort quite clearly: the ears will lie flat against his head; the tail may be held underneath the body; and it will be obvious that he is not enjoying the experience.

If he doesn't manage to wriggle free, and if his body language is ignored, he may try a threatening growl, bark or snap, which could lead to a bite if you're not careful. When a dog feels trapped in a small space, or by people surrounding him, he can quickly become aggressive; possibly liable to snap.

Time to stop?

Whether it is your grooming or something else that has upset him, you will need to work on this behaviour. If you're wondering whether to pay attention to his warnings and stop what you're doing, then the answer is definitely yes. It is not worth the risk of being bitten.

Try changing your approach so that it's less stressful for him. Take things in small steps: try brushing him gently for a short time, and then reward him with treats if he stays still and lets you proceed.

At the first signs of discomfort, let

A sudden movement has made this dog feel threatened, so she displays defensive behaviour in response.

him go. Gradually increase the time you spend grooming him, to let him get used to it all.

In the here and now

Dogs live in the here and now. Your dog will merrily greet you in the morning, forgetting that last night he sulked because no one would play with him. So if he happens to be in a bad mood, he should recover pretty easily if you distract him with a nice walk outdoors for example, or simply leave him in peace for a little while.

Exercise

Exercise will help relieve your dog's stress – be he in an aggressive mood or simply feeling a little out of sorts – warming his muscles and easing physical tension. Go for a good long walk in woods or cross country; if possible, make this an everyday outing, but don't try to involve him in any training games whilst you're out.

Leave him in peace

Try to calm the situation if he's being boisterous or distracted, or when he is racing around the living room, sniffing and barking. Help him to calm down by talking quietly and soothingly to him; perhaps trying a little gentle massage, then send him to his basket. Reward him with a treat if he stays there. Make sure all other family members leave him in peace. Give your dog some time out and he will soon come round.

Exercise helps to ease tension, uses up excess energy, and promotes a feeling of well-being for you and your dog!

Get your dog used to physical contact and being held by you, right from puppyhood.

dog speak

There's certainly no need for an interpreter when a dog barks; the situation and the dog's body language explain what he's saying quite clearly. However, if he barks when he's alone, this is usually a sign of anxiety.

Barking can have a variety of meanings, ranging from 'I'm excited!' to 'Keep your distance!'

Come and play with me!
Your dog is barking at you, looking you straight in the eye. He sits in front of you, or runs around you, and barks and barks and barks. You can guess exactly what he's trying to say: 'Give me my ball!' or 'Let me out into the garden!' If you want this commotion to stop, just ignore it. At first he will keep barking, infuriated that he is being ignored. But eventually he'll realise that he's not getting anywhere, and give up, though this may take a little while.

Dogs bark for many different reasons: it might be an invitation to play, but could also be a sign of frustration and boredom.

As soon as he stops making all that noise, give him your attention; throw him the ball or open the door for him.

I'm bored!
Dogs get bored, too. They might start barking at the garden fence or gate, not necessarily aggressively, but the more often they do it, the longer it will continue each time. If he is left on his own, he will not stop.

Keep your dog busy, and try to wear him out physically and mentally by making his walks more interesting with

sniffing games which are great for tiring energetic dogs (see *Smellorama: nose games for dogs,* published by Hubble and Hattie). You might even play some

doggie sports with him: agility jumps, frisbee, obedience games, even dog dancing – there are plenty of options which will not only be fun for your dog, but for you as well! (see *Dog Games: stimulating games to entertain your dog and you*, published by Hubble and Hattie.)

I'm scared

If your dog is barking because he feels uneasy, fearful, nervous or panicked, he will not stop, even when you ignore him and no matter how worn out he is. In this case, see if you can figure out what is causing his distress. If he feels uncomfortable about something, remove whatever it is that is irritating him. Perhaps you need to ask guests in your home not to crowd around him.

Confidently and quickly remove him from the situation that seems to be upsetting him. Some dogs with anxiety problems may require professional help.

My territory!

Lots of dogs bark from behind the garden fence, saying 'Go away! This is my territory!' Protected by this barrier, a dog feels more powerful, and he can bother other dogs without fear of the consequences. See page 48 for how to deal with this problem.

All on his lonesome

Barking when alone is an indication of stress. This type of bark will not be such a loud protest, and if it turns into howling, it means your dog is trying to gather together his family pack.

The only way to tackle this is to reduce the time he spends alone to shorter periods, with more distractions to occupy him.

'Stay away!' This dog feels threatened and is demanding his own space.

Try these tips:

• Leave your dog with a toy that contains a treat inside it: a Kong® for example; this will keep him busy!
• Try not to make your greetings and goodbyes too drawn-out or dramatic
• Limit your dog's space to just one room while you are out. He should not be allowed to roam over the entire house when he is alone. But always remember that he will need his basket, and, most importantly, water, though not necessarily food if it is not his mealtime.

Paws for thought:
'Human Speak'

Command signals

When training your dog, it's a good idea to repeatedly use one short word, like 'Go!' for example, which will act as a 'green light' signal. Then, whenever you say 'Go!', he knows it's time for action: not a request but an obligatory command which he must follow.

Speechless

Of course, your dog doesn't actually understand what you say to him, since his brain has no speech and language functions. However, he can learn to recognise a variety of tones in your voice, and to associate them with particular behaviours: 'Sit!' for example, prompting him to sit down.

However, it's sometimes the best policy to say nothing to your dog. Why? Because your dog could easily misunderstand a stream of words from you. If, for example, you said to your barking dog "Stay still and be quiet!", to him, this just sounds like you are joining in with his barking. Similarly, if your dog is showing signs of anxiety and nerves in an unfamiliar situation, you might try to reassure him by saying, "It's okay, sweetie, keep calm, we can go home soon." Unfortunately, your dog won't understand these words, and take them as confirmation that there is definitely something to be worried about, concluding: "This is definitely a strange situation and the human is worried about it, too, so I need to say on guard for danger."

So, when in doubt, keep quiet.

Your dog will only be able to recognise a few human words, and it will be a painstaking process for him to learn these. So use your body language as well as words, to make things easier for him

Working together

When training your dog, or playing a game with him, always reinforce and reward good behaviour with praise. Keep your tone warm and cheerful when you praise him, and emphasise your words with the type of communication he can actually taste – ie treats. Is this bribery? Only from our

point of view! Treats are your chance to reward his good behaviour. And if he is rewarded for something, he will do it again. You could go a step further and try clicker training: using a clicker and a dog treat together is a very clear way of communicating, which can quickly improve training sessions with your dog.

Don't do that!

A sharply spoken 'No!' or 'Stop!' are signal words to which your dog will react. But only use them in a clear voice, and don't use them too often, otherwise they will just become a part of the torrent of words your dog hears every day. Body language, on the other hand, can never be used too often. Stay conscious of your physical movements when communicating with your dog.

If you are not happy with something he's doing, express this clearly with your body language, by walking away

from him, for instance. Your dog will instantly understand, and take note.

Your body language

To help you communicate with your dog, follow these canine etiquette tips:

- Stand slightly to one side of your dog, rather than directly in front of him
- Avert your gaze – don't look him straight in the eye
- Try and make yourself appear smaller; instead of towering over your dog, crouch down a little

These actions will help your dog to relax whilst you are training him, enabling him to focus on what you are asking him to do.

Show confidence

Don't agitate your dog by shouting loudly, or using other negative body language. The most effective technique you can use is to remain at ease, calm and in control, demonstrated by:

- Clear, single words for commands
- Clear body language
- A friendly voice
- Training him to perform a different task every day

Clear and simple: I'm the peacemaker!

We are accustomed to interpreting the behaviour of our fellow humans, but we also tend to anthropomorphise dogs' behaviour, concluding that they are not really so different from us. Wrong! Dogs are very different from people in many ways.

🐕 **When two dogs have a disagreement, the third will try to play the role of peacemaker**

Personal space

People need to maintain some personal space when standing in front of strangers – too close, and we'll begin feeling uncomfortable – a distance of around arm's length is usually about right. In the same way, dogs also prefer at some distance from each other. If a dog's lead is too short, or he finds himself surrounded by crowds of people or other dogs in close proximity, then he will feel stressed by this invasion of his personal space.

Avoiding conflict

If one dog stands very close to another, this will be interpreted as provocation, creating tension and possibly conflict. So what should a sensible third dog do? Step in! If he places himself between the two others, thereby separating them, the tension will usually evaporate. This doesn't always work, though, in which case the third dog may get caught up in a fight.

Jealous? Moi?

When two people share a lingering embrace or cosy up together on the sofa, then of course we interpret this as

The Mastiff and the Bordeaux dog lock eyes, whilst the little black dog attempts to keep the peace.

From a dog's point of view, there's no personal space between these two, and it's only a matter of time before there's conflict, unless they are separated.

love and a close, affectionate bond. To your dog, however, this behaviour may set off alarm bells.

Fearing that your personal space has been invaded, and that a conflict may arise because of this, your dog will try his best to stand between you and your loved one, to separate you. In doing so, he is simply attempting to avoid a fight, and keep the peace the only way he knows how. Jealousy, therefore, has nothing to do with it.

Shame on you!

In order for a dog to feel ashamed of his poor behaviour, he firstly needs to understand that he has done something wrong. Dogs do not share the morals and values of people. Although they may appear to express remorse, they

are actually using appeasement signals, having read our body language, and realised that we are angry with them.

Your dog won't understand that you are still cross about that piece of bread and butter he stole half an hour ago. If he greets you as normal, but you begin scolding him about the bread, this will only confuse him and harm the bond between you.

Braveheart

To show courage means to appear undaunted in doing something, even when the outcome is unknown, and success is far from guaranteed. In this sense, dogs are not terribly courageous. For them, it all boils down to motivation – for example, the hungrier a dog is, the more likely he is to take risks to get food.

dog speak

Help:
I feel stressed!

Dogs are not burdened by the pressures of our human lives: hurrying to be on time, trying to juggle family life and career, or meeting project deadlines. And yet, they can still suffer from stress.

In fact, many dogs will show obvious signs of stress: recognising and understanding these will make life easier for you and your dog.

A certain amount of stress is a normal part of everyday life for a dog. But, as with people, too much will result in illness

The biological function of stress

Dogs actually need some stress. In the wild, stress hormones give dogs a surge of energy for hunting and catching prey, as well as the strength for fight or flight when necessary. Stress is a natural part of every animal's healthy survival instinct.

The important thing is that any stress must be short-lived, and does not become a long-term condition. For people and dogs alike, chronic stress can lead to mental or physical illness.

As dogs are often stressed by adjusting to new situations or new people, they might struggle at first with training, as nerves can temporarily prevent them from learning new things.

Frequent urination, defecation, or scratching the neck can all be signs of stress.

Recognising stress

The signs of stress can sometimes be confused with anxiety, submission or appeasement signals. So it's important to consider each specific behaviour of your dog, in order to make an accurate assessment of the situation. Martine Nagel and Clarissa von Reinhardt have put together a list of stress symptoms in their book, *Canine Stress*:

- Nervousness, restlessness or over-reacting
- Wetting or soiling (includes urinating on themselves)
- Mounting – especially within mixed-sex groups of dogs
- Exaggerated grooming
- Knocking over objects and barking loudly

- Allergies, diarrhoea or vomiting
- Loss of appetite, or craving too much food
- Body odour and skin problems
- Panting, twitching, snapping around himself (eg at flies)
- Scratching at the throat, or other displacement activities

Lack of concentration

If nothing seems to be working during training, and your dog keeps averting his gaze, or turning his whole body away from you, then it's time for a break.

Perhaps take a short walk, or just give him some time out. Your dog obviously feels a little overwhelmed by his training. Restore his confidence by repeating some exercises that he's already good at, and only attempt a new training exercise a little at a time.

Everyday routines

Since stress can have many different causes, there's no single solution for it. In addition to giving your dog breaks during training, and showing patience when introducing a new exercise, it also helps to keep an eye on what happens in his everyday life.

Dogs love a familiar routine because it makes them feel secure. A change to this will cause them to feel stressed; for example, if you move house or go on holiday.

In times like these, make sure your dog maintains a strict daily routine as usual: ie regular meal times, walks, playtime, and plenty of undisturbed rest. This will go a long way toward reducing canine stress levels.

If you think your dog's stress is caused by too much excitement going

on around him, try to calm things down, or give your dog some quiet space away from the chaos.

Breathe deeply

Make sure *you* also take time out to relax: it could be that your dog is simply mirroring your own emotions ...

Too much ... too little

Mental overload is not the only cause of stress. A lack of intellectual stimulation can also cause dogs to exhibit symptoms of stress, such as licking the paws excessively. Ensuring your dog has something to do is a great therapy for him: even better if you join in, too!

The Australian Sheepdog has quite clearly turned his body away from his master. Does this signal his disinterest? No, in fact it indicates that his training is overwhelming him, and he is showing signs of stress.

Keep an eye out:
Ouch! That hurts!

You throw a ball and your dog runs after it, but suddenly he stops and holds up one paw. It could be that his muscles are just stiff, or he might have a sprain. He may hobble along for a bit, but then he'll recover and happily continue on his way.

Even so, keep an eye on him, and check him over as soon as you can in case of injury.

Dogs who are in pain can react aggressively.

If it hurts

How animals experience pain has long been discussed and disputed. Scientific research into pain thresholds has shown that dogs can tolerate pain far better than can people. Dogs who are in pain exhibit the following characteristic behaviours:

- They become very sensitive, irritable and aggressive
- They will not allow themselves to be examined

Of course, it depends on the type of illness or injury as to how a dog reacts and copes with it. When a dog first senses pain, he may react aggressively out of fear or anxiety. In this case, it is a good idea to use a muzzle with him while you check and treat his injuries. You don't know how he'll react to pain or shock – and he might snap at you because even your gentle touch could worsen his pain.

Remember: all living creatures will instinctively try to avoid any experiences they find painful.

Chronic pain

Your dog's behaviour will clearly show if he is in acute pain. He will constantly lick his wound/affected area, scratch himself, raise a sore leg, or whimper. However, some injuries can go unnoticed for quite a while, such as back problems, or gradually worsening pain in the joints. This may develop slowly, over time, forcing a dog to adapt his movements. If you suspect this, it's important to take your dog to see a vet or qualified animal physiotherapist. They will be able to diagnose the problem; for example, he may have hurt his leg, and is putting too much pressure on another to compensate and avoid aggravating his injury.

Getting on a bit

Dogs get old eventually, just like us, and unfortunately they, too, may suffer from age-related health problems. Painful arthritis in your older dog's hips could make him grumpy and irritable if you stroke a sensitive area, and he may struggle to get up after a snooze (see *My dog has arthritis ... but lives life to the full!* by Hubble and Hattie). He may no longer be able to jump into the car, and have to stand there waiting for his personal servant (that's you!) to lift him in.

How does pain affect dog speak?

If a dog is in pain or discomfort from a physical problem, he will not feel particularly sociable. Illness or injury will also interfere with a dog's body language. So when he meets another dog, there might be a few unexpected misunderstandings between the two animals.

His hearing and eyesight will probably deteriorate, too – he'll stop being able to hear your call, and he may jump in surprise when someone walks up behind him. He may begin to sleep more, perhaps even while you're preparing food in the kitchen, which would be unheard of in his younger days! Some dogs may also become a little confused and forgetful, so be patient with him in all cases.

Old dogs know their own minds, giving them an extra special charm.

dog speak

Don't panic!

If your dog is prone to fearfulness and panic, it may not be possible to deal with this successfully on your own. You may need professional help, and it could take a long time for his fear to subside

Panic should not be confused with fear. Dogs who are panicked will freeze. Uncertainty about a situation will slowly turn to fear, and then develop into panic if their fear is not allayed. It is almost impossible to calm a panicked dog. It may be possible to control a small dog, but what is the best way to calm and reassure a panicky Collie?

Seeking shelter

Many dogs are panicked by the sound of fireworks on New Year's Eve. Your dog may try to hide behind the sofa, or anywhere he can feel safer like under a table or even inside a cupboard. He will also try to make his body appear smaller, with his ears lying flat on his head and his tail tucked under, his back rounded and his legs pulled in. He may either stand motionless, or pace about in nervous circles and creep along the walls of the room. His saliva production may increase, causing him to drool. Whilst in this state, he will not eat, and cannot be tempted by even the tastiest treat. As soon as the cause of his panic has disappeared, his behaviour will revert to normal.

The memory of an elephant

Unfortunately, dogs remember these

A safe place to hide: dogs who are in a panic will try to hide. Don't try and encourage him out of his hiding place, as he may be put on the defensive and possibly even snap.

Try to take measures to ensure your dog won't harm you or others, by letting him be wherever he feels safe and comfortable, without disturbance of any kind. In this way, his fear will gradually subside and he will feel able to cope with the world again. Professional help could be useful in this respect.

Calming him

As with anxious or uneasy dogs, do not try to calm your panicky dog by speaking to him softly. From a dog's point of view, this will just reinforce the idea that he is in danger. Just stay quiet, breathe deeply, and make your body language relaxed. If you have one, efer to the training programme from your vet or animal trainer. Your pity and sympathy are pretty useless to your dog – he needs you to act!

Wide open eyes and a steady gaze clearly express this dog's fear and uncertainty.

dreaded experiences very well. As soon as a situation looks like it may unfold all over again, he will fall into a panic, which could be even worse than before. So, what should you do?

Firstly, try to remove the cause of his panic, if possible. Secondly, there are number of options available for treating the panic itself. Your dog's signs of panic will be similar to his behaviour when he's stressed. A good vet can prescribe stress-relieving medication for your dog, and a behavioural therapist can teach you techniques to calm your dog before he slips into a panic.

Since a panicky dog may be prone to bite (for instance, if you attempt to coax him out from under the table) you need to take this condition seriously.

Dog demons

Xenophobia – Fear of strangers is quite normal: dogs always greet new people with caution.

Claustrophobia – The fear of being shut in a small space, making some dogs afraid of getting into a car.

Agoraphobia – The fear of being in wide open public places. This can be seen particularly in young dogs, characterised by stiff, slow movements, with neck jutting forward as he walks. Getting him used to public places should be done gradually, over time.

dog speak

At a glance: Relaxing together

confident and friendly carer: you should be systematic, give clear signals, and be in charge in every situation. From time to time, do something fun for your dog: hide a treat in a mouse hole whilst out on a walk. Your dog will be glad to spend time with you and will remain by your side.

Duty calls
Sometimes, hanging out with your dog can become a chore. Feeding, going out in the rain, hoovering up dog hairs, or worse, his annoying habits or lack of progress in training may be a source

No contact
During certain phases of a dog's life (eg puberty), the bond between your dog and you may become strained. Maybe he won't look at you, or come to you, walking on ahead and using his body language to show that he is ignoring you.

Making contact
When you are walking your dog, call him to you, but not too loudly. If he ignores you, temporarily put him back on his lead, so that you can hold onto him in an emergency. Always be a

of irritation. Do your best not to get too wound up when this happens, and try to enjoy your time with him. Take a break from training, if necessary.

Positivity

Relax! There's no such thing as a dog who does everything perfectly, and if you make sure you are as relaxed as possible, everything will seem easier.

Always care for your dog properly, no matter how stressed you may be. Focus on him, and pay attention to his mood. Keep your body language calm and relaxed. Your positive vibes will quickly be absorbed by your dog! This positivity will keep him motivated, and he will enjoy spending time with you.

Strict routine

Whether you are feeding him or taking him for a walk, dogs love a familiar routine. Sticking to this routine gives

him a great quality of life, because it makes him feel secure and happy.

Having fun

Throwing him a ball, taking him for a walk, or training him to do something new will ensure your dog is using his body and brain to their full potential, at the same time strengthening your bond.

When you're playing a game together, your dog should be able to respond to basic commands like 'sit!' and 'stay!'; therefore, he needs to be properly trained. This will ensure your dog behaves himself during the game and that you remain in charge. Use a command word like 'end!' to signify that the game is over, and always make full use of your body language when you communicate with him!

dog speak

Further reading

Dog Games: stimulating play to entertain your dog and you
by Christiane Blenski. Hubble and Hattie (2011). ISBN: 9781845843328
Smellorama: nose games for dogs
by Viviane Theby. Hubble and Hattie (2010). ISBN: 9781845842932.
My dog has arthritis ... but lives life to the full!
by Gill Carrick. Hubble and Hattie (2012). ISBN: 9781845844189
Canine Body Language – a photographic guide: interpreting the native language of the domestic dog
by Brenda Aloff. First Stone (2009). ISBN: 9781929242351
Dog Language
by R Abrantes. Dogwise. ISBN: 978-0966048407
Dog Body Language Phrasebook: 100 ways to read their signals
by Trevor Warner. Thunder Bay Press (2007). ISBN: 9781592237098
Dog Talk: how to understand what your dog is 'saying'
by Kay Roberts (Kindle edition). Digital Products Mall (2010). ASIN: B0037Z6Z7M
Why Does My Dog Do That?
by Robin Glover and Caroline Spencer (2010). ISBN: 9780956763907
The Secret Language of Dogs: how to communicate effectively with your dog
by Heather Dunphy. Apple Press (2011). ISBN: 9781845434137
Understanding and Handling Dog Aggression
by Barbara Sykes. The Crowood Press Ltd (2001). ISBN: 9781861264626
Dog Secrets
by David Ryan. lulu.com (2010). ISBN: 9781445261591
How to Behave So Your Dog Behaves
by Sophia Yin. TFH Publications (2005). ISBN: 9780793805433
The Dog Owner's Manual: operating instructions, trouble-shooting tips, and advice on lifetime maintenance
by David Brunner. Quirk Books (2004). ISBN: 9781931686853
The Dog Whisperer: the gentle way to train your best friend by the man who speaks dog
by Graeme Sims. Headline (2009). ISBN: 9780755317004
On Talking Terms with Dogs: calming signals
by Turid Rugaas. First Stone (2009). ISBN: 9781929242368
A-Z of Dog Training and Behaviour
by Kay White and Patrick Holden. Ringpress Books Ltd (1999). ISBN: 9781860540943

Websites

http://www.doglistener.co.uk/language/
Helps you to listen to your dog and understand what he is trying to tell you

http://www.diamondsintheruff.com/diagrams.html
Photographs and diagrams of canine body language

http://dogs.about.com/od/dogtraining/tp/dogbodylanguage.htm
A useful website explaining canine body language as well as links to plenty of useful information

http://www.teachingpuppies.com/reading-your-puppies-body-language
Learn more about some basic signals your dog may be sending you, and how to interpret these signals in order to promote a healthy and happy dog and owner

http://www.dogwise.com/
A website for all-things canine!

http://www.aspcabehavior.org/articles/50/Canine-Body-Language.aspx
Dog body language is an elaborate and sophisticated system of non-verbal communication that, fortunately, we can learn to recognize and interpret

http://www.dogstardaily.com/training/body-language
The dog blog for body language and dog news!

http://www.dogskool.com/canine-body-language-chart.html
Get an idea of what to look for in your dog, and recognise what emotions he is showing

http://en.wikipedia.org/wiki/Dog_communication
A guide to canine communication

http://www.dogspelledforward.com/canine-body-postures/
Information, photos, and some very informative videos

http://www.dogloversdigest.com/
Dogs are talking, but are we listening?

http://www.dogbreedinfo.com/articles/speakingdog.htm
What is your dog saying? Use the information, photos, and videos to find out!

http://uk.pedigree.com/health-and-training/caring-for-your-dog/reading-your-dogs-body-language
One of the reasons dogs make such good pets is their ability to communicate with us. In fact, they're often better at understanding us than we are at reading their body language

dog speak

By the same author ...

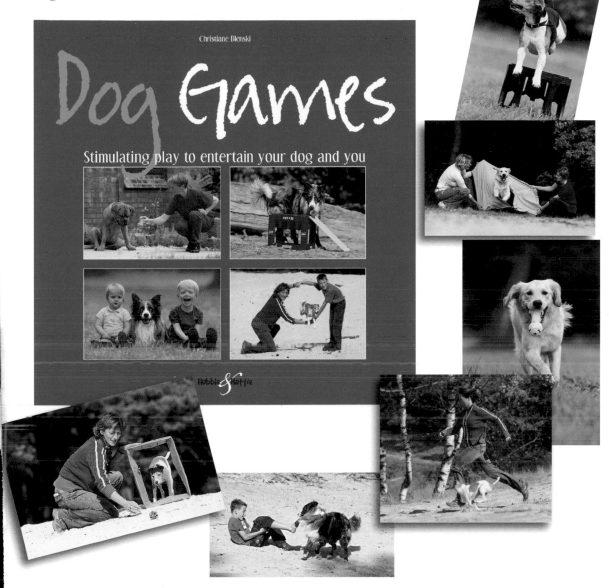

Christiane Blenski

Dog Games

Stimulating play to entertain your dog and you

Hubble & Hattie

Fantastic, new ideas for games that, after just a quick read of the instructions, allow you and your dog to get on with the fun business of playing!

Over 240 full-colour illustrations!

ISBN: 9781845843328
£15.99*

This unique book provides detailed observations of cat behaviour, and explains in everyday language how to interpret this. Essential reading for anyone who wants to better understand their feline friend

Over 130 full-colour illustrations!
ISBN 978184583854
£9.99*

For more info on Hubble and Hattie books visit www.hubbleandhattie.com; email info@hubbleandhattie.com; tel 44 (0)1305 260068. *prices subject to change; p&p extra